Living Mathematically

Probability In Our Lives

Mike Nerbovig

SCOTT Resources

Order Number: SR-1180
ISBN: 0-89212-228-5

About The Author

Mike Nerbovig lives in Eau Claire, Wisconsin with his wife Debbie and their two daughters. He is currently head of the mathematics department at Chippewa Falls High School in Chippewa Falls, Wisconsin where four of the mathematics courses offered were developed by him. They include: Mathematical Investigations which serves as an alternative course to pre calculus or as an enrichment class, Probability & Statistics which incorporates calculator simulations as the basis for exploration of these topics, Mathematical Research which involves an extension of calculus problems and investigation of diverse mathematical topics for students who have completed Advanced Placement Calculus, and Living Mathematically which is a year-long simulation intended to apply mathematical concepts which are essential for an individual to function successfully in our society.

Prior to teaching, Mike was an officer aboard a United States Navy destroyer and worked two summers as a research physicist at Los Alamos National Laboratory. His first teaching position was at a private school where he remained for seven years before leaving to teach overseas for an additional seven years with the U.S. Department of Defense schools.

In 1990 he was invited to be part of a new academic program for the United Arab Emirates University located in Al Ain, United Arab Emirates. That opportunity initiated him to curriculum writing and he built upon this experience to develop new curriculum when he returned to Chippewa Falls High School. He has been involved in the Wisconsin Academy Staff Development Initiative and has presented his work at a variety of workshops and conferences. His latest project is to complete his Survey of Mathematics course in hopes of exposing the power and beauty of mathematics to those students who have been disenfranchised from mathematics.

Living Mathematically Series

The last decade has been an exciting time to be involved in mathematics education. With expanded internet accessibility, graphing calculator technology, extensive interactive mathematics software, and new approaches to help students conceptualize mathematics, more opportunities exist than ever before for students to explore and understand the power and usefulness of mathematics.

Yet with all these scientific and educational advances, it still remains in the realm of the teacher to reveal the essence of mathematical thought and processes to students. Therefore, the curriculum and resources chosen by school districts should be aligned with their teachers' philosophies and the needs of their students.

The Living Mathematically modules were designed to engage at-risk students in significant mathematics as well as those students whose previous negative mathematical experiences jeopardize their success in future mathematics classes. These modules were adopted from the Living Mathematically curriculum which was developed around two specific components. The first component involves aspects of an ever changing society where financial transactions, planning for the future, and making informed decisions depend upon mathematical literacy. The second component exposes students to a world of mathematics usually not available in a traditional curriculum. From building projects to magic to chaos, students encounter the depth of mathematics and the way it influences their lives.

It is hoped that these Living Mathematically modules will unveil the excitement, fun, and relevance of mathematics to more students so that they and future generations of students become enthusiastic learners in the diverse world of mathematics.

Table of Contents

Introduction

Most students have acquired a rudimentary knowledge of probability by the time they reach high school. In fact, it would be rare indeed for a high school student not to be able to identify that the chances are 1 out of 2 of flipping a "heads" on a coin, or 1 out of 6 of rolling a "3" on a six-sided die. However, probability problems beyond those are not always so intuitive and often leave students frustrated and "turned off." Consequently, only a limited number of students who possess the analytical skills and motivation are usually able to experience the rich diversity of probability problems. By presenting different probability problems and concepts in this module in a non traditional context, it is hoped that students will realize that probability is comprehensible, and thus recognize the relevance and importance of probability in their lives.

Since probability is most often associated with gambling or guessing, the first activity asks students to guess numbers which will be spun by a spinner. Then students will investigate the mathematics of a familiar game show problem as well as a problem in disease diagnosis. Multiple-choice responses and an experiment in flipping a coin reveal expected and not so expected results. A ping pong game requires students to apply the appropriate mathematics to determine a reasonable payoff prize. Students next have to rank order different risks and decide whether certain probabilities are or are not acceptable for specific events. Finally two activities involving a spinner are used to demonstrate how much one can expect to win or lose in various games of chance.

Expectations

It is expected that students have previously been given the definition of probability and have an introductory knowledge of how to apply simple counting techniques in finding probabilities. Although minimal mathematics is required to complete the activities, probability values (where appropriate) are given as an instructional aid.

Organization

The activities in this module extend across a wide spectrum of probability problems, involve minimal materials, and engage students in critical mathematical thinking. Although this entire set of activities can serve as the basis for a probability unit, individual activities may also be incorporated as resource material for other presentations.

Activity 1: Guessing

BACKGROUND

There is evidence that the concept of probability or chance has existed since the beginning of recorded history. Games of chance played by tossing bones have been documented as early as 3000 B.C. Gambling with dice has been popular since Roman times, but was not studied mathematically until the sixteenth and seventeenth centuries. Today, probability is a useful tool in almost all fields of study.

For most people, however, the term probability conjures up images of slot machines, roulette tables, and the glitter of gambling casinos, as well as prospects of "striking it rich." Unfortunately, the dark side of gambling is not so well publicized. Consequently, the students will later be asked to explore how gambling adversely affects many individual's lives. But first the students will have the opportunity to play 4 games of lotto.

MATERIAL REQUIRED

Lotto Cards
Spinner
Student Activity 1 - 2 pages (1 copy per student)

TEACHER INFORMATION

Activity 1 – Use the blackline master entitled "Lotto Cards" to copy enough sheets so that each student has one. The students' first game of lotto is played by selecting four numbers between 1 and 36 and then filling in the appropriate circles on one of the four lotto cards. Then four numbers from 1 to 36 are to be randomly generated by use of the spinner card included in this module or by a calculator with such a capability. If the spinner card is used, then the outside ring of numbers is to be employed. If the spinner's arrow points to "0" or "00," then the spinner must be spun again. Likewise if a number is spun a second time or if the arrow falls on the line between two numbers, the spinner needs to be spun again. If any of the students match these four numbers in any order, they are declared a winner. This may also be the appropriate time to discuss the importance of the concept of randomness in such games of chance.

Allow the students three additional chances of winning by repeating the above procedure for the three remaining lotto cards.

Since the probability of winning this game is only 4/36 x 3/35 x 2/34 x 1/33 = 1/58,905, it is unlikely that any of the students would have won. The chances of winning this lotto game is just 1 out of 58,905 or stated another way; an individual would expect to win once if this game were played 58,905 times. To try to make this concept more relevant, the students will be asked to play some additional games where the probabilities of winning are much better.

Activity 1A – These games require use of the spinner card with the inside sectors numbered 1 through 10. As the directions are read from Student Activity 1, the students will be asked to play the games. When they have completed the games, they should have a better appreciation of the relationship between the probability of winning a game and the average (or expected) number of times that the game needs to be played in order to win.

After **Activity 1A** is completed, the students should be told that they would be expected to get the correct number 1 out of 10 trials because the probability of getting the correct number is 1 out of 10 each time the spinner is spun. Although individually the students may not have gotten just one correct number, the class average should be close to 1 correct out of 10.

After **Activity 1B** is completed, the students should be told that they would be expected to get the correct response for both spins 2.5 times out of 10 trials because the probability of identifying odd or even responses for both spins is 1 out of 4 or .25. Of course, the decimal expectation of 2.5 (from .25 x 10) might need further explanation, but the class average should be close to the theoretical value of 2.5 for the 10 trials.

Activity 1C – is designed to relate factual information to students while revealing the inherent dangers of gambling. The correct answers are listed below with some commentary. Hopefully, the students will think about the responses and become actively involved with a discussion as well.

Answers to Activity 1C:

1. True, in 1997 Maritz AmerPoll put the figure at 57%, and it continues to grow.
2. False, in 1999 The National Gambling Impact Study Commission cited surveys that listed the figure to be from 11.3% to 27.7%.
3. False
4. True
5. True
6. False, only a few schools have introduced such lessons in their mathematics curriculum.
7. True
8. False, the number is closer to 95%.
9. True
10. False, Americans wagered nearly $600 billion on lotteries, horse racing, casino gambling, and bingo in 1996, and the amount increases each year.
11. True, in fact it may be as high as 20 to 30 times as much.
12. True, and that is considering only legal gambling.
13. False, Australians are the heaviest gamblers in the world by a wide margin.
14. False, it is estimated that at least 5 people are affected for every person who gambles.
15. True

OPTIONAL STUDY

Powerball Lottery – Although there are many different lotteries, consider a popular type called powerball. To win, you have to pick 5 correct numbers out of 39 and then guess correctly the powerball from a choice of 39 numbers. The probability that you do that correctly is 5/39 x 4/38 x 3/37 x 2/36 x 1/35 x 1/39 = 1 out of 22,454,523. Based on the previous two activities, students should realize that they could be expected to win, on the average, once if they played this powerball game 22,454,523 times. At one game per day, that would be equivalent to more than 61,519 years.

Telephone Directories – Another way to look at this problem involves telephone directories. On a typical page there are approximately 300 listings of names. Because of the thin paper used in such directories, 100 pages (of 200 pages of names on both sides) constitute about 60,000 names at a thickness of approximately 1/4 inch. To have 22,454,523 distinct names would require about 374 directories of the this thickness (a total of almost 8 feet high). So winning the lottery is equivalent to having the correct 1 of these 374 directories randomly chosen which has your name in it. Then the correct side of 100 pages with your name on it would have to be randomly selected. Finally, if both of those conditions were met, your name out of 300 names would have to be randomly selected. With that minuscule probability, it almost seems like a miracle would have to occur for someone to win such a lottery!

Activity 2: Game Show Problem

BACKGROUND

Because of its counter intuitiveness, the famous "game show problem" of Marilyn vos Savant created considerable controversy as well as much discussion in the press several years ago. The problem involves a certain game show where the contestant is asked to select one of 3 doors. Behind one of the doors is a fabulous prize while the other two doors hide goats. Suppose the contestant guesses door number 1. Then, before revealing the contents of door 1, the game show host reveals either door 2 or door 3 to be hiding a goat and gives the contestant the opportunity to change his or her guess. The question is, of course, what is the best strategy for winning the prize.

MATERIALS

Scissors
Transparency of Game Show Door
Student Activity 2 (1 copy per student)

TEACHER INFORMATION

Discussion – Before investigating the problem, the students should reveal what strategy that they believe is best to win the prize. Then read the following information to the students.

Read to Class – In her regular column in *Parade* magazine, Marilyn vos Savant wrote that the best strategy for the contestant was to switch. Over the next few weeks, scores of mathematicians and scientists crucified Marilyn for "her mistake." Some of the comments that were reprinted in *Parade* are listed below.

> "You blew it, and you blew it big! I'll explain: After the host reveals a goat, you now have a one-in-two chance of being correct. Whether you change your answer or not, the odds are the same. There is enough mathematical illiteracy in this country, and we don't need the world's highest IQ propagating more. Shame!"
>
> "May I suggest that you obtain and refer to a standard textbook on probability before you try to answer a question of this type again?"
>
> "You are utterly incorrect about the game-show question, and I hope this controversy will call some public attention to the serious national crisis in mathematical education. If you can admit your error you will have contributed constructively toward the solution of a deplorable situation. How many irate mathematicians are needed to get you to change your mind?"
>
> "You are the goat."
>
> "You're wrong, but look at the positive side. If all those Ph.D.'s were wrong, the country would be in very serious trouble."
>
> "Maybe women look at math problems differently than men."

Before giving the students the correct answer (Marilyn vos Savant was right) the students should be given the opportunity to explore this problem by completing Student Activity 2.

Activity 2: Game Show Problem

Note: For many students the results from Student Activity 2 will not make it clear to them that the "switch" strategy is the most advantageous. However, you can make it clearer to them by providing a change of perspective. Instead of making the problem smaller, which is one technique that is often used in problem solving, making the problem significantly larger will demonstrate the preference of the "switch" strategy.

Teacher Directive – Use the blackline master entitled "100 Game Show Doors" to make a transparency. Do not turn the overhead projector on yet. Choose one of the 100 numbers. Then use an overhead marker to print "P" for prize over that number, and then cover that square with a small piece of a Post-it® note. Next tell the students that there is one fabulous prize behind one of the 100 closed doors. Then have a student choose one of the 100 doors. Similarly, cover that number with a small piece of a Post-it® note. Then turn the overhead projector on and tell the students that you uncovered 98 doors that had no prizes. Two unopened doors remain (covered with Post-it® notes), behind only one of which is the fabulous prize. Ask the student who picked the door number if he/she wants to switch or stay with their original choice. After the announcement, reveal the chosen door to show whether or not the student would have won the fabulous prize.

Most students should realize that the probability of winning the prize is not 1 out of 2 just because there are two doors left. It may take some discussion, but they should be able to understand that the probability of winning is 1 out of 100 with the "no switch" preference. However, it is an amazing 99 out of 100 with the "switch policy." Now is the time to review the original problem of three doors. Hopefully, with the analogy of the 100 doors, the students will realize that Marilyn vos Savant was correct and that the probability of winning by the switch policy is 2 out of 3 instead of the 1 out of 2 that most people believe is correct without investigating further. You may want to determine the class average to prove the point.

For those students who want to see the theoretical solution to the game show problem, the following is a mathematical explanation.

Consider three doors, numbered 1, 2, and 3. Behind one door is the prize. Since the contestant does not know which door it is, we are the liberty to choose any door. Let door 1 be our choice. With the use of four columns it will be shown that the switch strategy wins twice as often.

Contestant chooses	Game show hosts reveals	Contestant switches	Contestant does not switch
Door 1	Door 2 or 3 (makes no difference)	Loses	Wins
Door 2	Door 3	Wins	Loses
Door 3	Door 2	Wins	Loses

The contestant wins 2 out of 3 times by switching choices, but only 1 out of 3 by not switching choices.

Activity 3: Dreaded Disease

BACKGROUND

Another counter intuitive example involves disease diagnosis. After using Student Activity 3 to introduce the problem, have the students follow the directions in answering questions about this problem. Surprisingly the probability of actually having the disease after being tested positive is only 100 out of 1,098 or about 9% which is certainly not the overwhelmingly negative reaction that most of us would have had initially. In fact, this example should illustrate the need to investigate data beyond just what the numbers seem to indicate.

MATERIALS

Student Activity 3 (1 copy per student)

ANSWERS TO STUDENT ACTIVITY 3

a) How many people in this population have the disease? Calculate what 0.1% of 100,000 is.

0.1% • 100,000 = **100**

b) Of the number of people who have the disease, how many would test positive? Multiply the answer in part a) by 99%.

100 • .99 = **99**

c) Of the number of people not diseased, how many would test positive? Subtract the answer in part a) from 100,000. Then multiply this result by 1%.

99,900 • .01 = **999**

d) Add up the number of people who were tested to have the disease. Add up the answers in parts b) and c).

99 + **999** = **1098**

e) Now calculate the probability that you actually have the disease. Divide the answer in part a) by the answer in part d).

Probability that you have the disease = fraction = **9.1%**

Activity 3: Dreaded Disease

As an extension to the Dreaded Disease Activity, consider discussing the example which follows.

OPTIONAL STUDY

Two students in the same class are graded with 90%, 80%, 70%, and 60% as cut off percentages for grades of A, B, C, and D, respectively. The first student averaged 75% on his performance tasks and 85% on projects for a quarter. The second student, for the same grading period, averaged 74% on her performance tasks and 83% on projects. At the end of the quarter the first student received a "C" on his report card while the second student received a "B" on her report card. Since only the performance tasks and projects were used in determination of the quarter grade, how is it possible that the first student who scored higher on both performance tasks and projects could have gotten a lower grade than the second student? Here are the quarterly summaries of the two students.

	First Student		Second Student	
	Performance Tasks	Projects	Performance Tasks	Projects
	75	87	70	85
	80	81	68	88
	61		80	75
	77		63	90
	68		89	77
	82			
	81			
	72			
	84			
	70			
Averages	75	84	74	83

The answer can be found in how the teacher graded each student. The teacher offered the students the choice of two grading systems. The first system required completion of 10 performance tasks and 2 projects while the second system was based on the completion of 5 performance tasks and 5 projects. The teacher also weighted the projects as four times the weight of the performance tasks. Here are the quarterly summaries of the two students.

Using the weighted scales, the grades are determined below.

First student: $(75 \times 10 \times 1 + 84 \times 2 \times 4) \div 18 = 79$

Second student: $(74 \times 5 \times 1 + 83 \times 5 \times 4) \div 25 = 81.2$

Hence, the first student received a grade of "C" while the second student received a grade of a "B." Both the above example and the dreaded disease example show the need to investigate data whenever the results seem to contradict common sense.

Activity 4: Multiple-Choice Responses

BACKGROUND

Sometimes we base conclusions on exceptionally small probabilities. For example, in recent years the identification of an individual's genetic DNA has become so precise that criminals on "Death Row" have been proven innocent of their supposed crime because the probability of their DNA matching the crime scene's DNA is so remote that it is considered impossible. When certain rare events do occur, their small probabilities suggest that some other cause may have influenced the event. To illustrate how the effects of randomness and external factors can change probability values, two sets of multiple-choice responses will be used.

MATERIAL

Student Activity 4 - 2 pages (1 copy per student)

Activity 4A – This activity requires the students to work independently to fill in just one circle (a, b, or c) for each of 13 numbered items. However, before handing out this activity sheet, the instructor should fill in one circle for each of the 13 numbered items below so that these responses can later be compared to student responses.

1 ⓐⓑⓒ	6 ⓐⓑⓒ	11 ⓐⓑⓒ
2 ⓐⓑⓒ	7 ⓐⓑⓒ	12 ⓐⓑⓒ
3 ⓐⓑⓒ	8 ⓐⓑⓒ	13 ⓐⓑⓒ
4 ⓐⓑⓒ	9 ⓐⓑⓒ	
5 ⓐⓑⓒ	10 ⓐⓑⓒ	

Based on randomness, the probability that any one student matches a specific number of responses of the instructor is given in the chart below. Note: Using combinations, the first three computations are shown below.

$$p(0) = {}_{13}C_0 \cdot \left(\frac{2}{3}\right)^{13} = .0051, \quad p(1) = {}_{13}C_1 \cdot \left(\frac{1}{3}\right)\left(\frac{2}{3}\right)^{12} = .0334, \quad p(2) = {}_{13}C_2 \cdot \left(\frac{1}{3}\right)^{2}\left(\frac{2}{3}\right)^{11} = .1002$$

Number matched	Probability	Number matched	Probability
0	.0051	7	.0689
1	.0334	8	.0258
2	.1002	9	.00718
3	.1837	10	.00144
4	.2296	11	.000196
5	.2067	12	.0000163
6	.1378	13	.000000627

or about 1 in 1.6 million

After the students have been given the instructor's responses and determined the class average of number of matched responses with those of the instructor, this number should be compared with 4 which is the most probable number of matched responses as determined by theoretical considerations for randomness as seen in the chart above.

Additionally, the students can compare the numbers that they matched individually with the probabilities listed in the chart. Finally, some discussion should center around the randomness of the responses and the probability of matching all 13 responses.

Activity 4B – This activity, which is a variation to Activity 4A, illustrates how probabilities of events may be altered when responses are not random.

After the students have completed circling the 13 requested responses in the story, have each student fill in the top row of circles in the chart as follows. For the first response, tell the students to fill in circle A if 7:00 AM was circled on their paper, fill in circle B if 8:00 AM was circled, and fill in circle C if 9:00 AM was circled. This process should continue until a circle has been filled in for each of the 13 responses.

When this task is completed, one student will be asked to read slowly and distinctly each of his/her 13 responses while the remaining students fill in the appropriate circles in row 2. Then as each new student reads his/her responses, all the other students will fill in the circles for the responses in the next row.

When all the entries have been made, each student should count the number of matched responses of his/her row 1 with each other row and record that number to the far right of the chart. Finally, an average of the last column values should be computed to determine the average number of responses that the rest of the class made with each class member.

These number should vary and many should be higher than the number 4 of Activity 4A which was based on random responses. A discussion of this difference could be the basis for illustrating how multiple-choice responses are influenced by other factors than just randomness. In fact, examining each of the 13 responses separately might show some of the preferred responses for this particular group of students.

A discussion of additional multiple-choice responses could be included in this section. Since students often are asked to complete multiple-choice responses on surveys and achievement tests, references to these could be the basis for further discussions. Whenever there exist situations where multiple-choice responses are employed, probability computations will often accompany them so that appropriate inferences and conclusions can be made.

Activity 5: Flipping a Coin 200 Times

BACKGROUND
Another example which yields surprising results involves flipping a coin. Have the students write down in order what they believe might be the outcomes of flipping a coin 200 times in the allotted spaces for Day 1 on Student Activity 5. Then inform the students that their homework assignment is to flip a single coin 200 times at home and record, in order, whether the outcomes were "heads" or "tails." Then tomorrow you will hand back to the students their Student Activity 5 worksheets so they can copy their homework assignment information in the allotted spaces for Day 2.

Then these worksheets will be collected by you, and after a short examination, you will identify those students, if any, who did not perform the experiment, but rather just made up a list of "H's" and "T's" for Day 2. Because the probability of having a consecutive run of six or more heads or tails is extremely high, any student's list which does not have such a consecutive run has most likely faked the results. Although theoretically determining the probability of getting consecutive runs of six or more heads or tails is extremely difficult to compute, the results of actually performing the experiment are very revealing. This may require the students who have not already done so to flip a coin 200 times and record their results.

The discussion that follows should focus on the surprising results that there should be a run of at least six or more heads or tails out of the 200 trials. Since the probability of getting 1 head on 1 flip is 1/2, the probability of getting 2 consecutive heads on 2 flips is (1/2) (1/2), then the students should be able to recognize that the probability of getting 6 consecutive heads on 6 flips is (1/2) (1/2)(1/2) (1/2)(1/2) (1/2) or 1/64.

Therefore on average, a trial of 6 consecutive flips of a coin should result in 6 heads 1 out of 64 trials. It would also be true that 6 tails should also result once in 64 trials of flipping a coin 6 times. Thus it should not be so surprising that a run of at least six or more heads or tails would occur for flipping a coin 200 times.

In summary, it should be emphasized that with the proper knowledge of probability it is much easier to recognize false data as this example illustrates.

MATERIAL
Student Activity 5 (1 copy per student)
Coin

Activity 6: Insurance Costs

BACKGROUND

Perhaps no industry relies more on probability determinations than does the insurance field. From life insurance to disability insurance to car insurance, companies establish premiums based on projections from past statistical and probability calculations. Although the actual process of determining such rates is quite technical and extensive, **Student Activity 6** is intended to provide an overview of some of the considerations used in establishing insurance rates. In particular, students will be asked to determine the rates that individuals of varying ages should be charged for car and life insurance.

It should be stressed that the numbers and probabilities used in this activity are fictitious, and are only intended to be applied toward an analysis of the specifics of these exercises. However, an effort has been made to try to make the insurance costs realistic, especially across the different age groups.

MATERIAL

Student Activity 6 - 3 pages (1 copy per student)

Activity 6A – This activity which involves automobile insurance costs will probably be more relevant to the students than the second activity which involves life insurance costs. The first activity requires students to interpret data from a pictograph, determine the appropriate probability rates for automobile crashes per age group, and then calculate insurance rates. To illustrate, show the students that the probability rate per year for the 16-19 age group is approximately .156 (approximately 5.2 cars x 30 crashes per 1,000) while for the 45-54 age group it is approximately .048. Since the 16-19 age group is involved in crashes about 3.25 times more often than the 45-54 age group per 1,000 licensed drivers, a reasonable estimate of their insurance rate would be to multiply the 45-54 age group rate by 3.25 to get $1,137.50. Of course, these calculations did not take into consideration the severity of the crashes. Nonetheless, the idea is to demonstrate to the students how probability plays an essential role in so many aspects of our lives.

After this illustration, the students are asked to perform similar calculations, and then respond to a series of questions and statements about the probability rates.

Answers to Activity 6A:

Age Group	16-19	20-24	25-34	35-44	45-54	55-64	65-74	75-84	85+
P(crash) next year	.156	.099	.072	.057	.048	.042	.034	.037	.038

1. About $250
2. About $550
3. Older drivers are usually more experienced and cautious than younger drivers. But after 75 most people's reflexes slow down along with other age related impairments.
4. Hopefully this question will generate a good class discussion.

Activity 6B – This activity is designed to introduce students to some of the basic concepts used in determining whole life insurance for individuals. Before applying the assumptions of the activity, show the students that an individual at age 20 would have to make 600 monthly payments (at 0% interest) of $166.67 to accrue $100,000 by age 70. Actually the monthly value is $166.666 ...., but it becomes $166.67 when rounded to the nearest cent.

Next have the students consider 100 individuals of similar health, each 20 years of age, that want to be insured for a $100,000 whole life insurance policy to be paid up by the time they reach 70 years of age. Using the assumption from Student Activity 6B challenge the students to find a monthly payment at 0% interest for these 100 individuals if it is expected that exactly 50 will die at age 45 while the other 50 people live at least to age 70. The appropriate mathematics is shown below.

> $100,000 x 100 people = $10,000,000 that has to be paid to the policy holders.
> 50 people pay the monthly premium for 25 years or 300 months.
> The other 50 people pay the monthly premium for 50 years or 600 months.
>
> The monthly premium should be $10,000,000 ÷ (50 x 300 + 50 x 600) = $317.46

A more difficult challenge is to have the students redo the calculations for the previous problem where it is now expected that of the 100 individuals 10 will die at age 30, 20 will die at age 50, and the rest will live past 70 years of age. The appropriate mathematics is shown below.

> $100,000 x 100 people = $10,000,000 that has to be paid to the policy holders.
> 10 people pay the monthly premium for 10 years or 120 months.
> 20 people pay the monthly premium for 30 years or 360 months.
> 70 people pay the monthly premium for 50 years or 600 months.
>
> The monthly premium should be $10,000,000 ÷ (10 x 120 + 20 x 360 + 70 x 600) = $198.41

After discussing these examples with the students, direct them to complete **Activity 6B** including the problem that follows.

The appropriate mathematics for **Activity 6B** is shown below.

$100,000 x 100 people = $10,000,000 that has to be paid to the policy holders.
1 person pays the monthly premium for 5 years or 60 months.
1 person pays the monthly premium for 10 years or 120 months.
2 people pay the monthly premium for 15 years or 180 months.
2 people pay the monthly premium for 20 years or 240 months.
3 people pay the monthly premium for 25 years or 300 months.
4 people pay the monthly premium for 30 years or 360 months.
5 people pay the monthly premium for 35 years or 420 months.
6 people pay the monthly premium for 40 years or 480 months.
6 people pay the monthly premium for 45 years or 540 months.
70 people pay the monthly premium for 50 years or 600 months.

The monthly premium should be $10,000,000 ÷ (1 x 60 + 1 x 120 + 2 x 180 + 2 x 240 + 3 x 300 + 4 x 360 + 5 x 420 + 6 x 480 + 6 x 540 + 70 x 600) = $186.64

With all the assumptions made for the above assignment, it should be obvious to the students that the computations to determine actual insurance rates are quite involved and include numerous variables. To test their understanding of how changes would affect insurance rates, the students are also asked to place check marks appropriately for the last part of **Activity 6B**. The correct responses are listed below. Hopefully, the students might suggest how other changes might affect the determination of life insurance rates such as substituting term insurance for whole life insurance.

	increase or	decrease
If the policy value was $250,000 instead of $100,000.	☑	☐
If all the policy holders were women instead of men and women.	☐	☑
If the beginning age was 25 instead of 20 years of age.	☑	☐
If the P(death) was smaller for each age group until 70 years of age.	☐	☑
If the compounding interest rate were 3% instead of 0%.	☐	☑
If the policy holders had a medical history of heart disease.	☑	☐

BACKGROUND

Another approach in trying to develop student understanding of probabilities and their interpretations involves students who must establish parameters in a game setting. This activity invites students to investigate a game that is similar to one that is often played at carnivals and fairs. In this game thirteen identical bowls (approximately 6" in diameter) are arranged on a table top as shown in the diagram below.

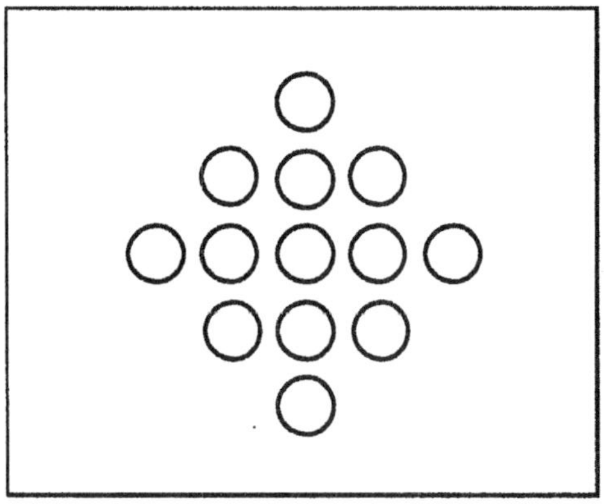

MATERIAL

13 Glasses/bowls (uniform in size) 6" diameter
Several Ping Pong balls
Student Activity 7 (1 per student)

TEACHER INFORMATION

The object of the game is to land a ping pong ball in one of the bowls by throwing it from a distance of several feet from the table. Glass bowls will increase the difficulty of achieving that objective while plastic bowls will decrease the difficulty. Since it is impossible to make a theoretical probability determination because of all the variables, an experimental probability will be found by performing a large number of trials. Therefore, as each student tosses a ping pong ball 50 times (or several ping pong balls for a total of 50 tosses), the number of times it ends up in any of the bowls should be recorded, as well as the number of times it ends up in the middle bowl.

If the ping pong balls are not removed once they have settled into a bowl, the chances are increased that another ping pong ball will end up in a bowl. Therefore, the ping balls should be removed to increase the difficulty of the game or left in place to decrease the difficulty of the game.

After the students record the number of ping pong balls that they succeeded in landing in the bowls, they should also record these values on the chalkboard or on an overhead projector which everyone will be able to view. Then the class should be guided through the calculations necessary to obtain a class average of the results.

The questions that then follow the activity are meant to make the students think more about the game, but more importantly, to make them apply probability concepts to answer the questions. The students should share their responses with the rest of the class to generate even more discussion.

Activity 8: Risk Assessment

BACKGROUND

Risk assessment can be defined as the process of assigning probabilities to harmful effects of human activities or natural disasters. However, risk assessment is often based on personal perceptions. For example, most people worry more about the dangers of flying from one destination to another even though the chances of getting into an accident are much less (by a factor of approximately 10) than those of driving for the same trip.

By using mathematics to analyze the probability of a risk, an individual can make an appropriate decision about whether or not the risk is worth taking. Of course, that decision is influenced by the experiences and personality of the individual. The same statement can be made about organizations and even governments.

When it was discovered that the probability of survival for ski victims after an avalanche decreased significantly after 15 minutes, a change in recommendations occurred. Instead of relying upon professional rescue teams as was done previously, mountaineering organizations are now suggesting that skiers should be taught techniques for finding, removing, and resuscitating avalanche victims. Similarly, when manufacturers of 3-wheel ATV's discovered that incidents of severe injuries resulting from accidents with these recreational vehicles had high probabilities, they voluntarily stopped their sales in the United States. And government regulations regarding cigarette sales and warning labels were instituted when the probability of developing lung cancer was identified as significantly greater for smokers than non smokers.

MATERIAL

Student Activity 8 - 4 pages (1 copy per student)

<u>Activity 8A</u> – The rationale behind the first part of Student Activity 8 is to allow students to discover the relative risks of a variety of activities rather than having them just presented factually. Therefore, the exercises ask students to rank order different risk activities or conditions, after which they are asked to discuss these risks when the actual probabilities associated with them are provided. The correct rank order and their corresponding probabilities are listed below.

ANSWERS: (Based on data provided in www.scottware.com.au)

8A – Dying in a sporting accident this year	
1. Mountain climbing	1 in 167
2. Motorcycling	1 in 1,000
3. Hang gliding	1 in 4,444
4. Running	1 in 10,000
5. Boating and swimming	1 in 36,000
6. Playing football	1 in 57,000

8B – Annual risk of dying from	
1. Cigarette smoking	3.6 out of 1,000
2. All cancers	2.8 out of 1,000
3. Motor vehicle accidents	.24 out of 1,000
4. Air pollution in eastern US	.20 out of 1,000
5. Home accidents	.11 out of 1,000

8C – Lifetime risk of	
1. Dying from heart disease	1 in 6
2. Committing suicide	1 in 71
3. Dying in a car accident	1 in 75
4. Being murdered	1 in 140
5. Being struck by lightning	1 in 10,456

8D – Another risk study shows loss of life expectancy (LLE) in days. Those rankings are:

1. Alcoholics	4,000
2. Living in poverty	3,500
3. Smoking	1,500
4. Being 20% overweight	1,000
5. Living near a nuclear plant	.5

The students should also be told that the probability of guessing the correct rank for all six activities in Activity 8A is 1 out of 720, and guessing the correct rank order for each of the three remaining rankings is 1 out of 120. Therefore, it is very likely that any student who correctly identified the rankings in order in any of the four categories has a good understanding of risks relative to the other activities in that category.

Activity 8E – The final exercise 8E involving risk assessment requires the students to shade in the appropriate region of probabilities that would be acceptable to them if they were participating in that designated activity. Note: the scale is logarithmic in order to display a greater range of probability values. Also the probability $10^{-6} = 1/10^6$ or is 1 out of 1,000,000. NOTE: If the student would not participate at any risk, then they should leave the region blank.

The following example could be shown to the students to demonstrate what region they would shade for an activity that they would participate in if the risk for serious injury were less than 1 out of 1,000.

Unshaded

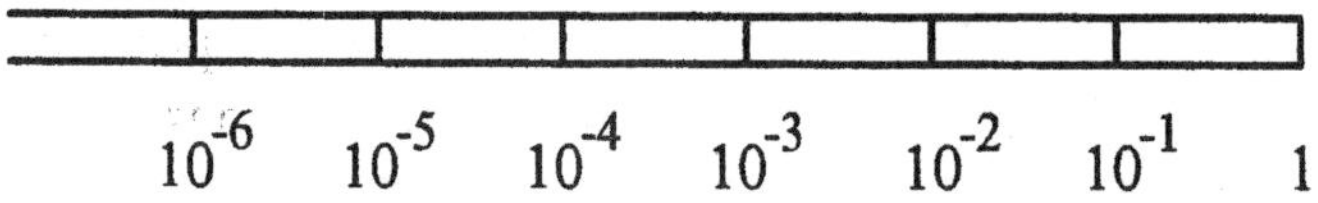

Shaded

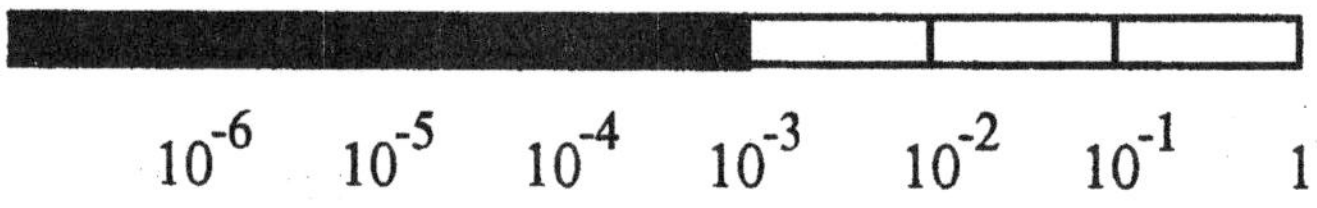

The students should now fill in the appropriate region for each category in 8E, after which a discussion should follow that invites students to explain their choices and compare their values to other members of the class.

Activity 9: Acceptable/Unacceptable Probabilities

BACKGROUND

Making decisions on whether or not a probability is acceptable to you often determines what course of action you might take. Unfortunately, students get very little guidance or practice in such decision making. Instead, most courses in probability focus on techniques to calculate various probabilities. The intention of this exercise is to help students make appropriate decisions based on the relative size of the corresponding probability. Of course such decisions depend upon the student's personality, which may be very conservative or particularly daring, and his/her value system. Regardless, **Student Activity 9** is an interesting exercise especially when individual student responses are compared with others in the class.

MATERIALS

Student Activity 9 (1 per student)

TEACHER INFORMATION

The extent of class discussions depend upon student and teacher interest. Some suggestions to facilitate a discussion are:

- Before you share your experiences and knowledge about some of the questions in this activity, the students should be given the opportunity to answer the questions individually, and then time should be allocated for students to discuss their responses with one another.
- Stress that there are no correct answers and the individual responses will vary depending upon personal experiences and life styles.
- Ask the students if they changed any of their responses or their reactions to them based on discussions with others in the class.

Additional comments regarding some of the questions are:

- Ask the students if their responses for questions 1 and 14 were the same. If they were not, then ask the students why they responded differently when the risk factor was the same.
- If students have never purchased a bag of apples as in question 2, they might not realize that getting at least one rotten apple is quite common.
- In reference to question 3, relate to students that a lot of people fear flying even when the probability of a plane crash is only 1 out of 1,000,000.
- In question 9 students will most likely not be aware of the high number of new vehicles that are recalled each year for a particular defect.

Activity 10: Games of Chance

BACKGROUND

For those individuals who become addicted to gambling, professional help is available and is probably necessary if they ever hope to cure themselves of this disease. However, early intervention might help thousands of others who do not recognize the potential dangers of adolescent gambling. That is why it is imperative that students be aware of the probability of losing and, in some instances, its severe consequences. Therefore, the final simulation of this module engages students in two probability activities in which bets are placed to illustrate the differences between the expectations of winning when no odds are provided (Student Activity 10A) and the expectations of winning when the odds are provided (Student Activity 10B).

MATERIALS

Student Activity 10 - 3 pages (1 copy per student)
Spinner
Chip/Coin (penny or dime)

<u>Student Activity 10A</u> – This activity requires the students to read a statement and then circle a "true" or "false" response for each of the trials. Next they will wager a specified amount by circling one of four choices. The instructor will then spin the spinner (included in this module) the appropriate number of times to determine how much was won or lost by each student for each trial. Because the probabilities of the event are not given in each trial, the students should be encouraged to apply their reasoning skills to determine whether or not the event is likely to happen. Hopefully students will recognize that if the probability is greater than .5, then betting with a "true" response would be in their favor. A second trial is provided to see if the students change their "true" or "false" response based on the outcome of the first trial. Their responses could be the basis for an interesting discussion on how outcomes of certain events might influence future decisions by people.

Another discussion could be prompted by asking the question: "Should the individual winnings or losses be attributed to luck or the individual's knowledge of whether or not it is likely that the event would happen?"

Note: In problem 5 it is necessary to compute the lengths of each of the three line segments and then recognize that the sum of the lengths of any two sides of a triangle must exceed the length of the remaining side. Also additional activity sheets may be used if more trials are desired.

<u>Student Activity 10B</u> – This activity simulates playing the game of "Roulette." Although a roulette wheel would be preferable, the spinner card from this module with the outer ring of numbers from 1 to 36 with "0" and "00" can be used. A description of the possible bets for this activity are included below and must be discussed with the students before the game can be played. Therefore, a transparency of the blackline master "Roulette Table" should be made. By placing a chip (or a penny as a substitute) on the overhead transparency, show the possible locations of the chip and the corresponding payoff if that number is spun or comes up on the roulette wheel.

To place a bet, a chip or chips must be placed appropriately on the roulette table. The highest probability of winning occurs for chips placed on the rectangular regions marked as: even, odd, dark oval, light oval, 1-18, or 19-36. See Figure A on the next page. For each chip bet, you get back the original plus one additional chip if the roulette ball ends on that designation. If it does not, then you lose the chip. For each chip placed on rectangular regions marked as: first column, second column, third column, first 12, second 12, or third 12 (see figure B below); you get back the original plus two additional chips if the roulette ball ends up on that designation. If it does not, then you lose the chip.

Other bets can be placed on any of six consecutive numbers having a common dividing line. A bet is made by placing the chip at the end of the line as shown in the figure C. If one of those numbers comes up on the roulette wheel, then five additional chips are paid to the bettor along with the return of the original chip. Similarly, bets can be placed on any four consecutive numbers by placing the chip on the common intersecting lines as shown in figure D. If one of those numbers comes up on the roulette wheel, then eight additional chips are paid to the bettor along with return of the original chip.

Finally, for each chip placed on any one of the 36 numbers or the "0" or "00" numbers, 35 chips are paid out plus the original one put down.

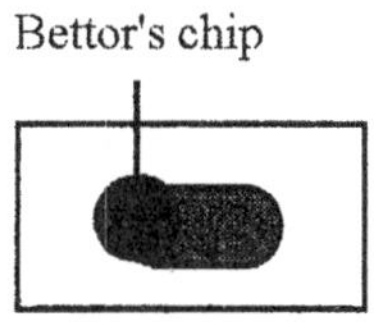

odds 1 to 1
Figure A

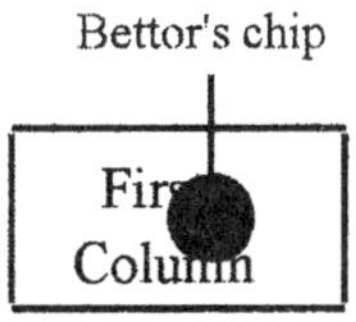

odds 2 to 1
Figure B

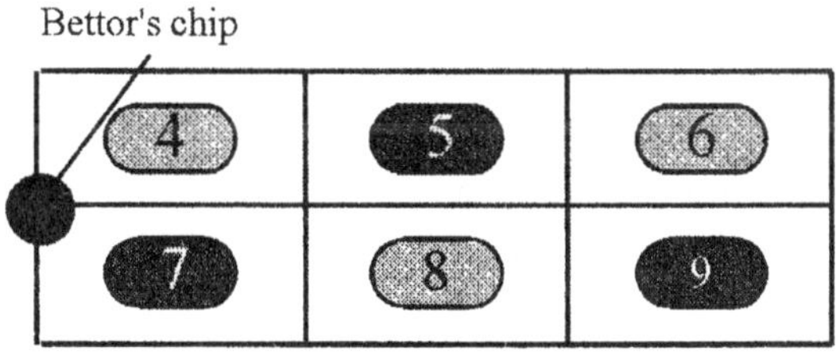

odds 5 to 1
Figure C

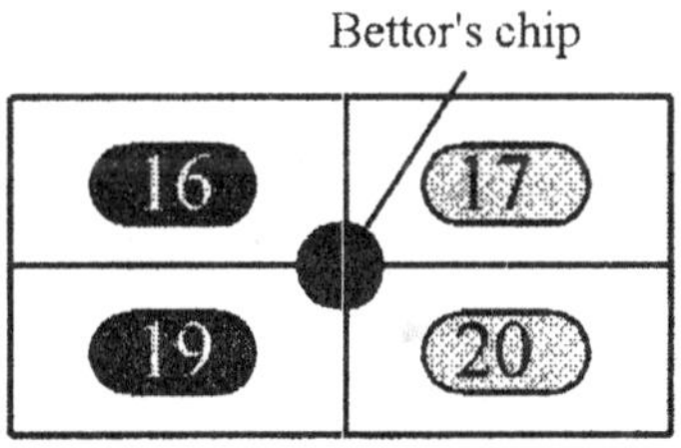

odds 8 to 1
Figure D

The odds for roulette are weighted in favor of the "house" (gambling establishment) which bases the expected profit margin on a large number of bets. All the odds are determined as if there were only 36 outcomes per roll of the roulette wheel or for the spinner for this simulation instead of the actual 38 outcomes. Thus there is a slight advantage for the house for each roll or spin.

For example, the probability of getting any of the 12 numbers in the first column is 12 out of 38, but the house pays out as though the probability were only 12 out of 36 or 1 out of 3. To show this house advantage, the use of an expectation = p(win) x (payoff) - p(losing) x (amount of bet) will be applied for the two situations where the payoff is \$2 for each \$1 bet.

No advantage: Expectation of the bettor = 12/36 x (\$2) - 24/36 x (\$1) = \$0

House advantage: Expectation of the bettor = 12/38 x (\$2) - 26/38 x (\$1) = - \$02/38

Because the better loses an average of 2/38 of a \$1.00 or approximately \$.05 on each \$1 bet on the first column of numbers, the house then expects to earn \$.05 for each such bet. The other house advantages can likewise be determined for the different possible bets in roulette.

Consequently, in this simulation it is expected that the total losses of the students will exceed the total winnings of the students. However, if enough trials are not performed (additional trials may be performed with additional activity sheets), the opposite is possible!

When all the trials are completed, the total winnings and losses of all the students should be calculated. Hopefully, enough trials were performed so that losses exceeded winnings. An appropriate follow up discussion should revolve around the fact that the winnings do exceed losses in real gambling situations. Of course, the winners are glamorized to attract even more gamblers while the losers are usually left alone to deal with their own personal losses.

STUDENT ACTIVITIES

Student Activity 1 *Name:*

Guessing

Activity 1A: Your instructor has a spinner with equal sectors numbered 1 through 10. You are to guess one of these numbers and fill in the appropriate circle for trial 1 below. Then your instructor will spin the spinner, after which you are to then fill in the appropriate circle to the right which indicates whether or not you matched the number spun by your instructor. This process will now be repeated nine additional times with the results recorded below.

	1	2	3	4	5	6	7	8	9	10	correct	incorrect
Trial 1	O	O	O	O	O	O	O	O	O	O	O	O
Trial 2	O	O	O	O	O	O	O	O	O	O	O	O
Trial 3	O	O	O	O	O	O	O	O	O	O	O	O
Trial 4	O	O	O	O	O	O	O	O	O	O	O	O
Trial 5	O	O	O	O	O	O	O	O	O	O	O	O
Trial 6	O	O	O	O	O	O	O	O	O	O	O	O
Trial 7	O	O	O	O	O	O	O	O	O	O	O	O
Trial 8	O	O	O	O	O	O	O	O	O	O	O	O
Trial 9	O	O	O	O	O	O	O	O	O	O	O	O
Trial 10	O	O	O	O	O	O	O	O	O	O	O	O
Total											____	____

After each member of the class gives the number that he/she got correct, you will determine and record the class average below.

The class average is ______ correct out of 10.

Activity 1B: Your instructor will again use the spinner with sectors numbered 1 through 10. This time you are to guess whether the first number spun is "odd" or "even" and whether the second number spun is "odd" or "even." For each of the ten trials fill in two circles below corresponding to your two choices, and then after your instructor reveals the numbers spun, record whether or not you matched both responses by filling the appropriate circle to the right.

	first spin odd	first spin even	second spin odd	second spin even	both correct	one or more incorrect
Trial 1	O	O	O	O	O	O
Trial 2	O	O	O	O	O	O
Trial 3	O	O	O	O	O	O
Trial 4	O	O	O	O	O	O
Trial 5	O	O	O	O	O	O
Trial 6	O	O	O	O	O	O
Trial 7	O	O	O	O	O	O
Trial 8	O	O	O	O	O	O
Trial 9	O	O	O	O	O	O
Trial 10	O	O	O	O	O	O
Total					____	____

After each member of the class gives the number that he/she got correct, you will determine and record the class average below.

The class average is ______ correct out of 10.

Activity 1C: Identify the following as true or false by circling T or F for each statement.

T F 1. More than 50% of Americans play the lottery.

T F 2. Less than 10% of adolescents show signs of being "problem" or "pathological" gamblers.

T F 3. Habitual gambling behavior among youth is not associated with drug and alcohol use.

T F 4. Legalized gambling increases participation by juveniles in all forms of gambling, both legal and illegal.

T F 5. Students with gambling problems are likely to sacrifice school, parents, and friends in order to continue gambling.

T F 6. At least half the schools in the U.S. offer a mathematics curriculum that includes lessons on the mathematics of gambling.

T F 7. Most adults put gambling pretty far down the list of concerns about kids.

T F 8. Approximately 65% of adolescent gamblers are boys.

T F 9. Gambling is the fastest-growing cause of personal bankruptcy in the U.S.

T F 10. Americans spend less than $100 billion annually on gambling.

T F 11. The amount of money bet illegally in the U.S. is more than 10 times the amount that is bet legally.

T F 12. Gambling revenues in the U.S. are more than movie revenues in the U.S.

T F 13. Americans are the heaviest gamblers in the world.

T F 14. Gambling only affects those people who gamble.

T F 15. Nearly 1 in 5 homeless people cite gambling as a cause of their situation.

Student Activity 2 Name:

Game Show Problem

Class set up: The class is to be divided into groups of two. If there is an odd number of students, the instructor can form the final pair with the remaining student.

Directions: Each student will cut out the rectangles at the bottom of this page for the 3 doors, 2 goats, and the fabulous prize. Each player will act as the game show host for twenty trials while the other group member will act as the contestant. The game show host will place the 3 doors in front of the other group member and then secretively place a goat or prize under each of the doors. The contestant then chooses one of the three doors, after which the host reveals a goat underneath one of the two remaining doors. The contestant will then be offered the choice of switching his original door selection or not. The host then reveals whether there is a goat or fabulous prize underneath the final chosen door. If a goat is under the selected door, then the host will record a tally mark under "goat" for the appropriate "switch" or "no switch" strategy. If a prize is under the selected door, then the host will record a tally mark under "prize." After 20 trials have been performed, the two players will reverse roles and 20 additional trials will be recorded by the new host.

Experimental Probability: Next each player will total up the "goats" and "prize" tally marks and determine the experimental probability identified at the bottom of the charts.

Switch Strategy

	Prize	Goat
Tally Marks		
Total		

P(Prize) = $\dfrac{\text{Total number of tally marks for prizes}}{\text{Total number of tally marks for prizes and goats}}$

P(prize) = ______

No Switch Strategy

	Prize	Goat
Tally Marks		
Total		

P(Prize) = $\dfrac{\text{Total number of tally marks for prizes}}{\text{Total number of tally marks for prizes and goats}}$

P(prize) = ______

Does there now appear to be a preference of "switch" or "no switch"?

Door 1	Door 2	Door 3	Goat
			Goat
			Fabulous Prize

Dreaded Disease

Suppose there is a test for a dreaded disease that is 99% accurate in the following sense. If someone has the disease, the test will be positive 99% of the time. However, if a person does not have the disease, the test will be negative 99% of the time and positive the other 1% of the time. Suppose further that 0.1% of the population actually has the disease.

1. Assume that you had just been tested positive for this dreaded disease. Write a sentence or two about what you believe would be your initial reaction to this alarming news.

2. After you had time to think about this situation, what would you do? Respond with a brief sentence or two.

3. Assume that the population for this problem is 100,000. Now perform the following calculations as directed below.

   a) How many people in this population have the disease? Calculate what 0.1% of 100,000 is.

   0.1% • 100,000 = ________

   b) Of the number of people who have the disease, how many would test positive? Multiply the answer in part a) by 99%.

   ________ • .99 = ________

   c) Of the number of people not diseased, how many would test positive? Subtract the answer in part a) from 100,000. Then multiply this result by 1%.

   __________ • .01 = ________

   d) Add up the number of people who were tested to have the disease. Add up the answers in parts b) and c).

   ________ + ________ = ________

   e) Now calculate the probability that you actually have the disease. Divide the answer in part a) by the answer in part d).

   Probability that you have the disease = fraction = __________

4. Now write a sentence or two about what your reaction would be to being tested positive for this dreaded disease .

Multiple-Choice Responses

Activity 4A: Below is a set of 13 multiple-choice items with 3 possible responses for each. Your task will be to fill in one circle for each of the 13 items to try to match the responses of your instructor who has previously filled in the responses for an identical set of items. Before starting this task, estimate below the number of responses that you believe you will match with those of your instructor's responses.

Number of responses that you believe you will match = _____

Now fill in the spaces.

1 ⓐⓑⓒ	6 ⓐⓑⓒ	11 ⓐⓑⓒ
2 ⓐⓑⓒ	7 ⓐⓑⓒ	12 ⓐⓑⓒ
3 ⓐⓑⓒ	8 ⓐⓑⓒ	13 ⓐⓑⓒ
4 ⓐⓑⓒ	9 ⓐⓑⓒ	
5 ⓐⓑⓒ	10 ⓐⓑⓒ	

As your instructor reads his/her responses, count the number of your responses that match the instructor's. Then as each student in class reads the number of matches that they counted, record the number below along with your number of matched responses.

_____, _____, _____, _____, _____, _____, _____, _____, _____, _____

_____, _____, _____, _____, _____, _____, _____, _____, _____, _____

_____, _____, _____, _____, _____, _____, _____, _____, _____, _____

Now find the average number of matched responses by you and your classmates and record below.

Average number of matched responses is ________

Multiple-Choice Responses

Activity 4B: Read the following story. When you encounter a set of parentheses, circle the response that best suits you.

Yesterday must have been my lucky day. When I woke up Saturday morning at (7:00 AM, 8:00 AM, 9:00 AM), there was a message for me to call a car dealership. When I called the dealer, she told me that I had just won a brand new (Ford Focus, Chevrolet Cavalier, Honda Civic). After I picked up my new vehicle, I rode to my favorite fast food restaurant (Mc Donalds, Burger King, Hardee's) and discovered that I was the one millionth customer so my lunch was free. While eating my lunch I saw a fifty dollar bill lying on the floor. I immediately picked it up and (gave it to the store manager, put it in the Multiple Sclerosis donation container, put it in my pocket). Then I went to a local dairy to get (one, two, three) scoops of (chocolate, vanilla, strawberry) ice cream. Since it was celebrating many years of business in the community, I got the ice cream cone free.

With the extra money I decided to get my hair cut and colored. I thought it might be interesting to see how people would react if I changed my hair color to (purple, red, green). The hair stylist was going to beauty school so he was cutting, styling, and coloring hair for free. Can you believe my luck? I felt so generous, I tipped him ($1, $2, $5). Just then the most beautiful person that I ever saw walked into the shop. We started talking about my new hair color and agreed to go out to dinner for (steak, seafood, fast food). It seemed that we talked endlessly about (sports, music, school). Suddenly it was (9:00 PM, 10:00 PM, 11:00 PM) and time to go home. Before leaving we made arrangements to go to a movie the next night which would be (an adventure, a romance, science fiction). When I got home, my parents were (asleep, watching TV, waiting up to talk to me). As I jumped into bed, I hoped that I would not wake up to realize that the luckiest day of my life had just been a dream.

When all the students have completed the above task, each student will be asked to read his/her responses so that they can be entered into the chart on the next page as described by your instructor.

Activity 4B - Cont'd.

Student	1. ABC	2. ABC	3. ABC	4. ABC	5. ABC	6. ABC	7. ABC	8. ABC	9. ABC	10. ABC	11. ABC	12. ABC	13. ABC	Number Matched
1	OOO	OOO	OOO	OOO	OOO	OOO	OOO	OOO	OOO	OOO	OOO	OOO	OOO	1 X
2	OOO	OOO	OOO	OOO	OOO	OOO	OOO	OOO	OOO	OOO	OOO	OOO	OOO	2 ___
3	OOO	OOO	OOO	OOO	OOO	OOO	OOO	OOO	OOO	OOO	OOO	OOO	OOO	3 ___
4	OOO	OOO	OOO	OOO	OOO	OOO	OOO	OOO	OOO	OOO	OOO	OOO	OOO	4 ___
5	OOO	OOO	OOO	OOO	OOO	OOO	OOO	OOO	OOO	OOO	OOO	OOO	OOO	5 ___
6	OOO	OOO	OOO	OOO	OOO	OOO	OOO	OOO	OOO	OOO	OOO	OOO	OOO	6 ___
7	OOO	OOO	OOO	OOO	OOO	OOO	OOO	OOO	OOO	OOO	OOO	OOO	OOO	7 ___
8	OOO	OOO	OOO	OOO	OOO	OOO	OOO	OOO	OOO	OOO	OOO	OOO	OOO	8 ___
9	OOO	OOO	OOO	OOO	OOO	OOO	OOO	OOO	OOO	OOO	OOO	OOO	OOO	9 ___
10	OOO	OOO	OOO	OOO	OOO	OOO	OOO	OOO	OOO	OOO	OOO	OOO	OOO	10 ___
11	OOO	OOO	OOO	OOO	OOO	OOO	OOO	OOO	OOO	OOO	OOO	OOO	OOO	11 ___
12	OOO	OOO	OOO	OOO	OOO	OOO	OOO	OOO	OOO	OOO	OOO	OOO	OOO	12 ___
13	OOO	OOO	OOO	OOO	OOO	OOO	OOO	OOO	OOO	OOO	OOO	OOO	OOO	13 ___
14	OOO	OOO	OOO	OOO	OOO	OOO	OOO	OOO	OOO	OOO	OOO	OOO	OOO	14 ___
15	OOO	OOO	OOO	OOO	OOO	OOO	OOO	OOO	OOO	OOO	OOO	OOO	OOO	15 ___
16	OOO	OOO	OOO	OOO	OOO	OOO	OOO	OOO	OOO	OOO	OOO	OOO	OOO	16 ___
17	OOO	OOO	OOO	OOO	OOO	OOO	OOO	OOO	OOO	OOO	OOO	OOO	OOO	17 ___
18	OOO	OOO	OOO	OOO	OOO	OOO	OOO	OOO	OOO	OOO	OOO	OOO	OOO	18 ___
19	OOO	OOO	OOO	OOO	OOO	OOO	OOO	OOO	OOO	OOO	OOO	OOO	OOO	19 ___
20	OOO	OOO	OOO	OOO	OOO	OOO	OOO	OOO	OOO	OOO	OOO	OOO	OOO	20 ___
21	OOO	OOO	OOO	OOO	OOO	OOO	OOO	OOO	OOO	OOO	OOO	OOO	OOO	21 ___
22	OOO	OOO	OOO	OOO	OOO	OOO	OOO	OOO	OOO	OOO	OOO	OOO	OOO	22 ___
23	OOO	OOO	OOO	OOO	OOO	OOO	OOO	OOO	OOO	OOO	OOO	OOO	OOO	23 ___
24	OOO	OOO	OOO	OOO	OOO	OOO	OOO	OOO	OOO	OOO	OOO	OOO	OOO	24 ___
25	OOO	OOO	OOO	OOO	OOO	OOO	OOO	OOO	OOO	OOO	OOO	OOO	OOO	25 ___
26	OOO	OOO	OOO	OOO	OOO	OOO	OOO	OOO	OOO	OOO	OOO	OOO	OOO	26 ___
27	OOO	OOO	OOO	OOO	OOO	OOO	OOO	OOO	OOO	OOO	OOO	OOO	OOO	27 ___
28	OOO	OOO	OOO	OOO	OOO	OOO	OOO	OOO	OOO	OOO	OOO	OOO	OOO	28 ___
29	OOO	OOO	OOO	OOO	OOO	OOO	OOO	OOO	OOO	OOO	OOO	OOO	OOO	29 ___
30	OOO	OOO	OOO	OOO	OOO	OOO	OOO	OOO	OOO	OOO	OOO	OOO	OOO	30 ___

Now find the average number of questions guessed correctly by you and your classmates and record below.

Average number of questions guessed correctly ____________

Flipping a Coin 200 Times

Day 1: Imagine you were given a homework assignment of flipping a coin 200 times and recording the outcomes in order. However, you forgot to do it. So just before class you fake the results by writing down a series of H's and T's on your paper and hand it in to your instructor. Pretend this happened to you, and write down your series of H's and T's below.

Day 2: Record neatly below your homework results of flipping a coin 200 times. Use H for heads and T for tails. Remember, the results must be recorded in the order that they occurred in the experiment.

When completed, turn in this activity sheet to your instructor.

Your instructor will now discuss the differences, if any, between the data from Day 1 and Day 2.

Student Activity 6 Name:

Insurance Costs

Activity 6A: Below is a pictograph of *Automobile Crashes by Age Group.*

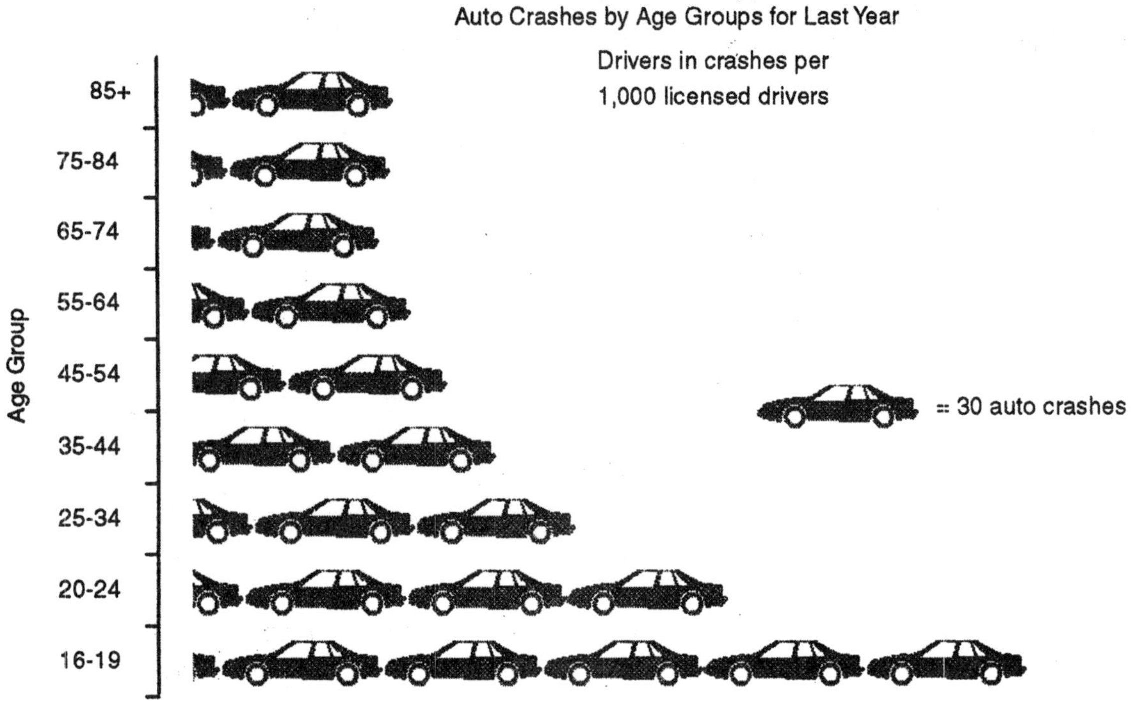

If 45-54 year old drivers were paying $350 per year on the average for insurance coverage, what would be a reasonable estimate for comparable insurance coverage for 16-19 year old drivers? Record your estimate below.

16-19 year old drivers estimated yearly insurance cost = $ ___________

After comparing your values with others in the class and discussing the reasons for their estimates, you might be convinced that determining what insurance rates should be is much more complicated than just comparing crashes for age groups. In fact this task is performed by actuaries (mathematicians) that are employed by insurance companies.

To see how probabilities can be used to help in such a determination, record the approximate probability rates for automobile crashes per age group from the pictograph in the chart below.

Age Group	16-19	20-24	25-34	35-44	45-54	55-64	65-74	75-84	85+
P(crash) next year									

Using this chart, respond to the questions or statements on page 2 of this activity.

Activity 6A Cont'd.

1. If the insurance rate for 45-54 year old drivers were $350 per year, what would be a reasonable estimate for comparable insurance coverage for 65-74 year old drivers?

2. If the insurance rate for 25-34 year old drivers were $250 per year, what would be a reasonable estimate for comparable insurance coverage for 16-19 year old drivers?

3. Give a possible explanation as why the number of crashes steadily decline as the drivers get older until the age of 65-74.

4. Do you think it is fair that insurance companies generally offer a 10-20% discount to honor roll students? Why?

Activity 6B: This activity is designed to introduce you to some of the basic concepts used in determining whole life insurance for individuals. Some assumptions will be applied to make this exercise easier to complete. They are:

1) There will be a 0% compounding rate for all monthly premiums throughout the life of the insurance policy.
2) The monthly installment rates for each age category will remain constant.
3) The life insurance policies will be paid monthly until the age of 70 when they will be fully paid.
4) The values given in the chart below do not differentiate between male and female policy holders.
5) All the policy holders are in similar, good health.
6) The deaths of insured policy holders will be at the 5-year increments: 25, 30, 35, etc.
7) There are 100 insured policy holders, each at 20 years of age.
8) The insured value of each policy holder will be $100,000.

The following chart identifies the estimated probability that an individual will die at one of the five-year increments.

Age	25	30	35	40	45	50	55	60	65	70+
P(death)	.01	.01	.02	.02	.03	.04	.05	.06	.06	.70

Note: Before beginning the assignment, follow the example(s) provided by the instructor.

Your assignment is to find a monthly payment at 0% interest for these 100 individuals that will provide each with a $100,000 benefit under the assumptions listed above. Show the appropriate mathematics in the space below.

Activity 6B Cont'd.

With all the assumptions made for the above assignment, it should be obvious that the computations to determine actual insurance rates are quite involved and include numerous variables. Rather than make such calculations, you are asked to identify if the monthly rates would increase or decrease according to the changes listed below by placing a check mark appropriately.

If the policy value was $250,000 instead of $100,000.	☐	☐
If all the policy holders were women instead of men and women.	☐	☐
If the beginning age was 25 instead of 20 years of age.	☐	☐
If the P(death) was smaller for each age group until 70 years of age.	☐	☐
If the compounding interest rate were 3% instead of 0%.	☐	☐
If the policy holders had a medical history of heart disease.	☐	☐

Student Activity 7 *Name:*

Ping Pong Game

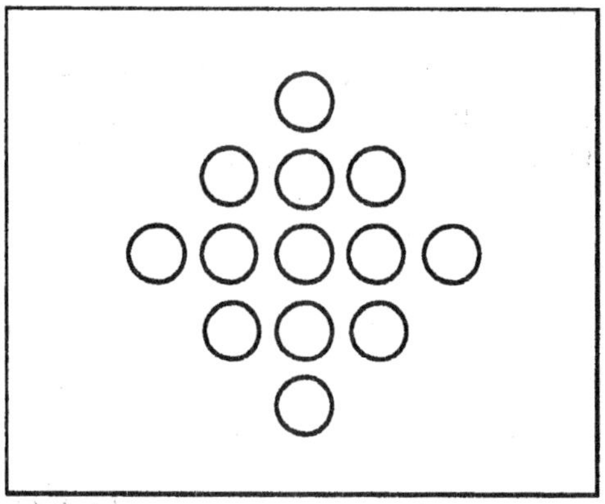

The object of the game is to land a ping pong ball in one of thirteen bowls placed in a pattern on a table top as shown in the diagram above. Your instructor will locate the positions from which you may throw the ping pong balls. Students should be paired up so that as one student tosses a ping pong ball 50 times, the other student can record below the number of times it ends up in any of the bowls and the number of times it ends up in the middle bowl.

You made _____ out of 50 in all the bowls and _____ out of 50 in the middle bowl

Your partner made _____ out of 50 in all the bowls and _____ out of 50 in the middle bowl

Class average of these numbers:

_____ out of 50 in all the bowls and _____ out of 50 in the middle bowl

Questions: Based on your observations for the game just completed and the class average just computed, respond to each of the questions or statements below.

1. If it costs $1.00 for tossing three ping pong balls, what should be the value of the prize given for landing a ping pong ball in a bowl at the break even value?

2. If you were in charge of running this game, what do you think would be a reasonable value for the prize in problem 1 if you hoped to make some money from this game?

3. Let us assume that you wanted to offer a "super prize" for landing a ping pong ball in the middle bowl and smaller prizes for landing a ping pong ball in any of the smaller bowls. What would be your values for each of these prizes? Why?

4. Do you think additional practice would enable you to land the ping pong ball in the bowls more often?

5. Do you think carnival personnel take mathematics into consideration when they devise these types of games of chance?

Activity 8A: Rank order from 1 (highest) to 6 (lowest) those activities below which have the highest probability of resulting in death this year for the participants.

	Rank Order						Correct Order
boating and swimming	①	②	③	④	⑤	⑥	______
hang gliding	①	②	③	④	⑤	⑥	______
mountain climbing	①	②	③	④	⑤	⑥	______
motor cycling	①	②	③	④	⑤	⑥	______
playing football	①	②	③	④	⑤	⑥	______
running	①	②	③	④	⑤	⑥	______

Number that you had in the correct order is ______

Activity 8B: Rank order from 1 (highest) to 5 (lowest) those situations below which have the highest probability of resulting in death this year for the participants.

	Rank Order					Correct Order
air pollution in the eastern U.S.	①	②	③	④	⑤	______
all cancers	①	②	③	④	⑤	______
cigarette smoking	①	②	③	④	⑤	______
home accidents	①	②	③	④	⑤	______
motor vehicle accidents	①	②	③	④	⑤	______

Number that you had in the correct order is ______

Student Activity 8 *Name:*

Activity 8C: Rank order from 1 (highest) to 5 (lowest) those situations below which have the highest probability of a lifetime risk for the participants.

	Rank Order					Correct Order
being murdered	①	②	③	④	⑤	______
being struck by lightning	①	②	③	④	⑤	______
committing suicide	①	②	③	④	⑤	______
dying in a car accident	①	②	③	④	⑤	______
dying from heart disease	①	②	③	④	⑤	______

Number that you had in the correct order is _______

Activity 8D: Rank order from 1 (highest) to 5 (lowest) those conditions below which have the highest loss of life expectancy (LLE) in days for individual participants.

	Rank Order					Correct Order
alcoholics	①	②	③	④	⑤	______
being 20% overweight	①	②	③	④	⑤	______
smoking	①	②	③	④	⑤	______
live near nuclear plant	①	②	③	④	⑤	______
living in poverty	①	②	③	④	⑤	______

Number that you had in the correct order is _______

Activity 8E: Fill in the appropriate region as discussed by your instructor for the probabilities that would be acceptable to you if you were participating in that designated activity. NOTE: If you would not participate in this activity at any risk, leave the region blank.

1. Playing your favorite sport with the risk that you would break your leg from falling.

10^{-6} 10^{-5} 10^{-4} 10^{-3} 10^{-2} 10^{-1} 1

2. Speeding on the interstate with the risk of being issued a $250 ticket.

10^{-6} 10^{-5} 10^{-4} 10^{-3} 10^{-2} 10^{-1} 1

3. Investing $1,000 for a return of $10,000 one year later with the risk of losing the $1,000 investment.

10^{-6} 10^{-5} 10^{-4} 10^{-3} 10^{-2} 10^{-1} 1

4. Working at a biological laboratory for $50/hour with the risk of developing cancer.

10^{-6} 10^{-5} 10^{-4} 10^{-3} 10^{-2} 10^{-1} 1

5. Flying in a friend's airplane with the risk of crashing.

10^{-6} 10^{-5} 10^{-4} 10^{-3} 10^{-2} 10^{-1} 1

6. Having a baby with the risk that the baby would have a permanent crippling disability.

10^{-6} 10^{-5} 10^{-4} 10^{-3} 10^{-2} 10^{-1} 1

7. Living on a spectacular hillside with the risk of a mudslide washing away your home.

10^{-6} 10^{-5} 10^{-4} 10^{-3} 10^{-2} 10^{-1} 1

Activity 8E Cont'd.

8. Visiting a foreign country with the risk of a terrorist attack against you.

10^{-6} 10^{-5} 10^{-4} 10^{-3} 10^{-2} 10^{-1} 1

9. Attending a party with the risk of being beaten up.

10^{-6} 10^{-5} 10^{-4} 10^{-3} 10^{-2} 10^{-1} 1

10. Going on an African safari with the risk of contracting malaria.

10^{-6} 10^{-5} 10^{-4} 10^{-3} 10^{-2} 10^{-1} 1

11. Living in a neighborhood with the risk of having your home burglarized.

10^{-6} 10^{-5} 10^{-4} 10^{-3} 10^{-2} 10^{-1} 1

12. Bungy jumping with the risk of dying.

10^{-6} 10^{-5} 10^{-4} 10^{-3} 10^{-2} 10^{-1} 1

Student Activity 9　　　*Name:*

Acceptable/Unacceptable Probabilities

Understanding probabilities is extremely important. Not only do such probabilities determine actions by companies and even governments, but they affect our daily lives as well. This exercise requires interpretation of probabilities to answer the questions. Circle "yes" or "no" for each question asked. Afterwards, you will be asked to discuss your responses with another member of your class to see if you are in agreement with one another. Then your answers will be compared to those of your instructor.

1. If the probability that your friend gets in a car accident is 1 out of 10, would you ride with that friend? yes no

2. If the probability that you will get a rotten apple in a bag is 1 out of 40, would you buy that bag of apples? yes no

3. If the probability that a plane crashes is 1 out of 1,000, would you fly on that plane? yes no

4. If the probability that your Friday night date will be gorgeous is 1 out of 5, would you go on that blind date? yes no

5. If the probability that a new eye contact lens will cause blindness is 1 out of 2,000, would you buy this contact lens? yes no

6. If the probability of rain is 2 out of 3, would you go to the swimming pool? yes no

7. If the probability that you will be fined for speeding is 1 out of 8, would you speed? yes no

8. If the probability that your favorite team will win its next game is 1 out of 3, would you go to the game? yes no

9. If the probability that a new car would be recalled for a defect is 1 out of 50, would you buy that new car? yes no

10. If the probability that you run out of gas is 1 out of 10, would you drive around town in winter? yes no

11. If the probability were 1 out of 2 that you could live an extra 10 healthy years by drastically changing your eating habits, would you change your eating habits? yes no

12. If the probability that a major asteroid will crash into the earth during your lifetime is 1 out of 200, would that concern you? yes no

13. If the probability that your 2-year old sister wanders from the front yard into the street is 1 out of 4, would you leave her unattended in the front yard? yes no

14. If the probability that your friend gets in a car accident after drinking alcohol is 1 out of 10, would you ride with that person after he/she drank alcohol? yes no

Student Activity 10 *Name:*

Activity 10A: After reading each of the problems below, you are to circle "true" or "false" in reference to the statements provided. Then you may bet $0, $10, $50, or $100 by circling the appropriate amount. After your instructor performs each trial, you are to record the amount that you won or lost in the designated space. When your instructor announces that it is time to quit, you are to add up our total winnings or losses and record at the bottom of the page.

1. The sum of two randomly generated integers from 1 to 10 exceeds 10.

Trial 1:	true or false	Bet: $ 0 $10 $50 $100
Trial 2:	true or false	Bet: $ 0 $10 $50 $100

won _______ or lost _______

2. The second of two randomly generated integers from 1 to 10 is greater than the first.

Trial 1:	true or false	Bet: $ 0 $10 $50 $100
Trial 2:	true or false	Bet: $ 0 $10 $50 $100

won _______ or lost _______

3. The sum of two randomly generated integers from 1 to 10 is even.

Trial 1:	true or false	Bet: $ 0 $10 $50 $100
Trial 2:	true or false	Bet: $ 0 $10 $50 $100

won _______ or lost _______

4. You will be two or more steps from your starting position after randomly generating four integers from 1 to 10 if you walk one step to the right if you get an even integer and one step to the left for an odd integer.

Trial 1:	true or false	Bet: $ 0 $10 $50 $100
Trial 2:	true or false	Bet: $ 0 $10 $50 $100

won _______ or lost _______

5. A number line starting at 0 units and ending at 11 units is divided into three line segments by two points which are determined by randomly generating two different integers from 1 to 10. The resulting three line segments will join to form a triangle.

Trial 1:	true or false	Bet: $ 0 $10 $50 $100
Trial 2:	true or false	Bet: $ 0 $10 $50 $100

won _______ or lost _______

Games of Chance

Activity 10B: After a discussion of the game of roulette, you will be given an opportunity of playing it with simulated money. You will be allowed to continue placing bets for each spin of the spinner until you have lost $100 or more. For each spin, record the category of numbers from the list below that you are betting on. After your instructor performs each spin you are to record the amount that you won or lost in the designated space. Note: If you win, the amount that you bet is multiplied by the given factor for each spin and recorded as your winnings. If you lose, then the amount that you bet is what is lost. When your instructor announces that it is time to quit, you are to add up our total winnings or losses and record at the bottom of the page.

Categories:		
	1) even, odd, dark oval, light oval, 1 - 18, 19 - 36	win = 1 x bet
	2) 1st 12, 2nd 12, 3rd 12, 1st column, 2nd column, 3rd column	win = 2 x bet
	3) any array of six numbers	win = 5 x bet
	4) any array of four numbers	win = 8 x bet
	5) any single number including "0" and "00"	win = 35 x bet

Spin # 1: Category numbers: ____________________ amount of bet = $ ________
amount won = $ _______ amount lost = $ ________
cumulative: winnings = $ _______ losses = $ ________

Spin # 2: Category numbers: ____________________ amount of bet = $ ________
amount won = $ _______ amount lost = $ ________
cumulative: winnings = $ _______ losses = $ ________

Spin # 3: Category numbers: ____________________ amount of bet = $ ________
amount won = $ _______ amount lost = $ ________
cumulative: winnings = $ _______ losses = $ ________

Spin # 4: Category numbers: ____________________ amount of bet = $ ________
amount won = $ _______ amount lost = $ ________
cumulative: winnings = $ _______ losses = $ ________

Spin # 5: Category numbers: ____________________ amount of bet = $ ________
amount won = $ _______ amount lost = $ ________
cumulative: winnings = $ _______ losses = $ ________

Activity 10B Cont'd.

Roulette

Spin # ____: Category numbers: ____________________ amount of bet = $ ________
amount won = $ _______ amount lost = $ ________
cumulative: winnings = $ _______ losses = $ ________

Spin # ____: Category numbers: ____________________ amount of bet = $ ________
amount won = $ _______ amount lost = $ ________
cumulative: winnings = $ _______ losses = $ ________

Spin # ____: Category numbers: ____________________ amount of bet = $ ________
amount won = $ _______ amount lost = $ ________
cumulative: winnings = $ _______ losses = $ ________

Spin # ____: Category numbers: ____________________ amount of bet = $ ________
amount won = $ _______ amount lost = $ ________
cumulative: winnings = $ _______ losses = $ ________

Spin # ____: Category numbers: ____________________ amount of bet = $ ________
amount won = $ _______ amount lost = $ ________
cumulative: winnings = $ _______ losses = $ ________

Spin # ____: Category numbers: ____________________ amount of bet = $ ________
amount won = $ _______ amount lost = $ ________
cumulative: winnings = $ _______ losses = $ ________

Spin #____: Category numbers: ____________________ amount of bet = $ ________
amount won = $ _______ amount lost = $ ________
cumulative: winnings = $ _______ losses = $ ________

Spin # ____: Category numbers: ____________________ amount of bet = $ ________
amount won = $ _______ amount lost = $ ________
cumulative: winnings = $ _______ losses = $ ________

Spin # ____: Category numbers: ____________________ amount of bet = $ ________
amount won = $ _______ amount lost = $ ________
cumulative: winnings = $ _______ losses = $ ________

Spin # ____: Category numbers: ____________________ amount of bet = $ ________
amount won = $ _______ amount lost = $ ________
cumulative: winnings = $ _______ losses = $ ________

Blackline Masters

Lotto Cards

Living Mathematically Lotto Cards

Select any four of the numbers from 1 through 36 by filling in the appropriate circles.

First Number		Second Number		Third Number		Fourth Number	
⓪	⓪	⓪	⓪	⓪	⓪	⓪	⓪
①	①	①	①	①	①	①	①
②	②	②	②	②	②	②	②
③	③	③	③	③	③	③	③
	④		④		④		④
	⑤		⑤		⑤		⑤
	⑥		⑥		⑥		⑥
	⑦		⑦		⑦		⑦
	⑧		⑧		⑧		⑧
	⑨		⑨		⑨		⑨

Living Mathematically Lotto Cards

Select any four of the numbers from 1 through 36 by filling in the appropriate circles.

Living Mathematically Lotto Cards

Select any four of the numbers from 1 through 36 by filling in the appropriate circles.

Living Mathematically Lotto Cards

Select any four of the numbers from 1 through 36 by filling in the appropriate circles.

100 Game Show Doors

1	2	3	4	5	6	7	8	9	10
11	12	13	14	15	16	17	18	19	20
21	22	23	24	25	26	27	28	29	30
31	32	33	34	35	36	37	38	39	40
41	42	43	44	45	46	47	48	49	50
51	52	53	54	55	56	57	58	59	60
61	62	63	64	65	66	67	68	69	70
71	72	73	74	75	76	77	78	79	80
81	82	83	84	85	86	87	88	89	90
91	92	93	94	95	96	97	98	99	100

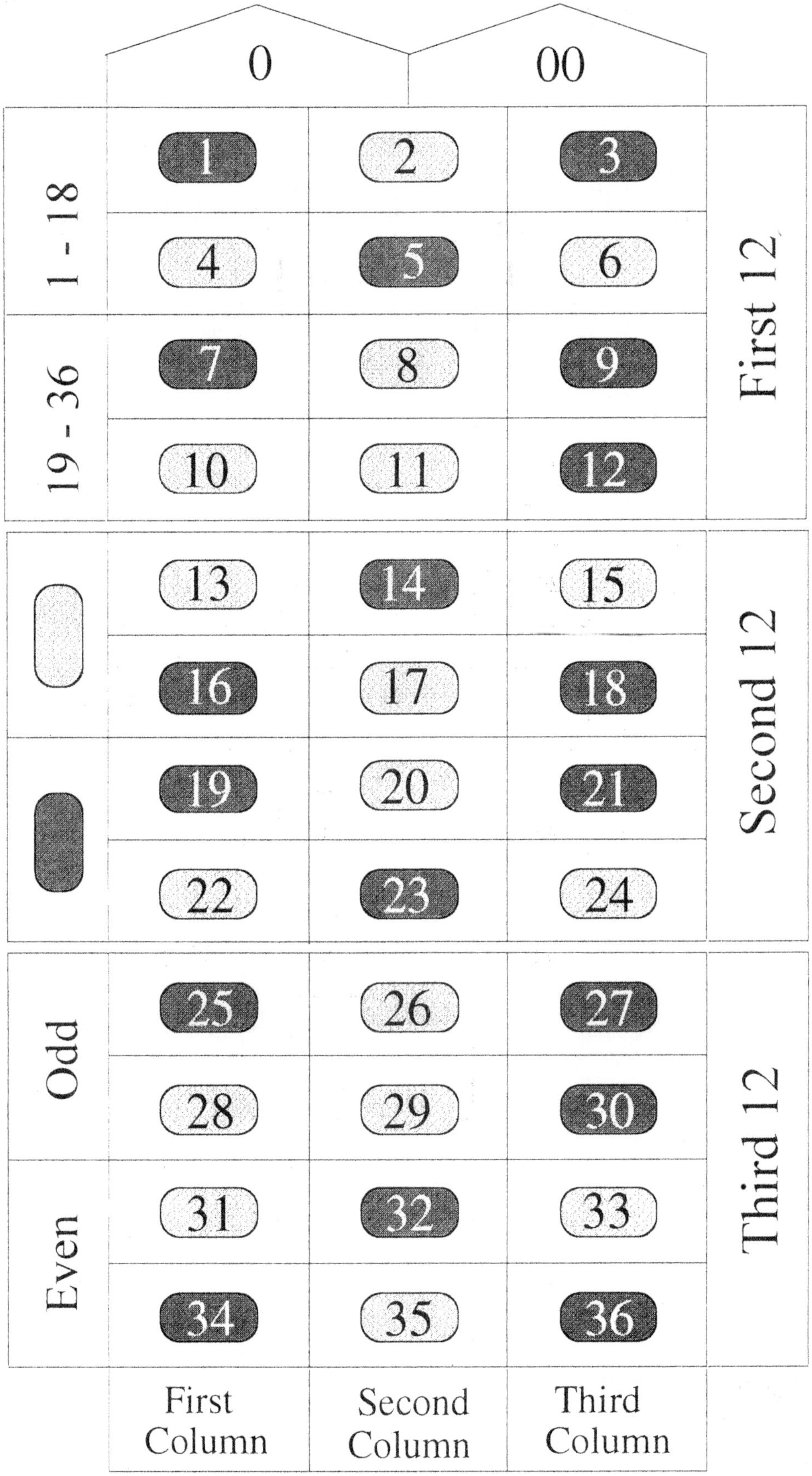
0
00
1 - 18
19 - 36
Odd
Even
1
2
3
4
5
6
7
8
9
10
11
12
13
14
15
16
17
18
19
20
21
22
23
24
25
26
27
28
29
30
31
32
33
34
35
36
First 12
Second 12
Third 12
First Column
Second Column
Third Column

CORRELATION TO THE NCTM STANDARDS

GRADES 9-12

These standards are based on the recommendations made in 1988 by the Board of Directors of the National Council of Teachers Of Mathematics in their publication. Curriculum and Evaluation Standards for School Mathematics. The standards included here are those specifically targeted for high school students

Standards	1	2	3	4	5	6	7	8	9	10
Standard 1: Mathematics as Problem Solving										
Use problem solving approaches to investigate and understand mathematical content.		X	X	X	X	X	X	X		X
Apply integrated mathematical problem-solving strategies to solve problems from within and outside mathematics.						X				
Recognize and formulate problems from situations within and outside mathematics.			X	X	X	X	X	X	X	X
Apply the process of mathematical modeling to real world problem situations.	X	X	X	X		X	X			X
Standard 2: Mathematics as Communication										
Reflect upon and clarify their thinking about mathematical ideas and relationships.	X	X	X	X	X	X	X	X	X	X
Formulate mathematical definitions and express generalizations discovered through investigations.		X	X							
Express mathematical ideas orally and in writing.	X		X	X	X	X	X		X	
Read written presentations of mathematics with understanding.										
Ask clarifying and extending questions related to mathematics they have read or heard about.		X	X	X		X			X	
Appreciate the economy, power, and elegance of mathematical notation and it's role in the development of mathematical ideas.										X
Standard 3: Mathematics as Reasoning										
Make and test conjectures.		X	X			X				
Formulate counter examples.									X	
Follow logical arguments.		X	X	X						
Judge the validity of arguments.						X			X	
Construct simple valid arguments.			X		X				X	
Construct proofs for mathematical assertions, including indirect proofs and proofs by mathematical induction.										
Standard 4: Mathematical Connections										
Recognize equivalent representations of the same concept.									X	
Relate procedures in one representation to procedures in an equivalent representation.		X		X						
Use and value the connections among mathematical topics.										
Use and value the connections between mathematics and other disciplines.	X			X						X
Standard 5: Algebra										
Represent situations that involve variable quantities with expressions, equations, inequalities, and matrices.										
Use tables and graphs as tools to interpret expressions, equations, and inequalities.										
Operate on expressions and matrices, and solve equations and inequalities.										
Appreciate the power of mathematical abstraction and symbolism.										
Use matrices to solve linear systems.										
Demonstrate technical facility with algebraic transformations, including techniques based on the theory of equations.										
Standard 6: Functions										
Model real-world phenomena with a variety of functions.										
Represent and analyze relationships using table, verbal rules, equations, and graphs.		X					X			
Translate among tabular, symbolic, and graphical representation of functions.										
Recognize that a variety of problem situations can be modeled by the same type of function.										
Analyze the effects of parameter changes on the graphs of functions.										
Understand operations on, and the general properties and behavior of, classes and functions										

Standard 7: Geometry form a Synthetic Prospective										
Interpret and draw three-dimensional objects.										
Represent problems situations with geometric models and apply properties of figures.										
Classify figures in terms of congruence and similarity and apply these relationships.										
Deduce properties of, and relationships between, figures from given assumptions.										
Develop and understanding of an axiomatic system through investigating and comparing various geometries.										
Standard 8: Geometry from an Algebraic Perspective										
Translate between synthetic and coordinate representation.										
Deduce properties of figures using transformations and using coordinates.										
Identify congruent and similar figure using transformations.										
Analyze properties of Euclidean transformations and relate translations to vectors.										
Deduce properties of figures using vectors.										
Apply transformations, coordinates, and vectors in problem solving.										
Standard 9: Trigonometry										
Apply trigonometry to problem situations involving triangles.										
Explore periodic real-world phenomena using the sine and cosine functions.										
Understand the connection between trigonometric and circular functions.										
Use circular functions to model periodic real-world phenomena.										
Apply general graphing techniques to trigonometric identities.										
Solve trigonometric equations and verify trigonometric identities.										
Understand the connections between trigonometric functions and polar coordinates, complex numbers, and series.										
Standard 10: Statistics										
Construct and draw inferences from charts, tables, and graphs that summarize data form real-world situations.		X				X		X		
Use curve fitting to predict from data.										
Understand and apply measures of central tendency, variability , and correlation.	X			X						
Understand sampling and recognize its role in statistical claims.										
Design a statistical experiment to study a problem, conduct the experiment, and interpret and communicate the outcomes.		X			X		X			
Analyze the effects of data transformations on measures of central tendency and variability.										
Transform data to aid in data interpretation and prediction.										
Test hypotheses using appropriate statistics.										
Standard 11: Probability										
Use experimental or theoretical probability, as appropriate, to represent and solve problems involving uncertainty.	X	X	X	X	X	X	X	X	X	X
Use simulations to estimate probabilities.	X	X		X			X			X
Understand the concept of a random variable.				X						X
Create and interpret discrete probability distribution.										
Describe the normal curve and use its properties to answer questions about sets of data that are assumed to be normally distributed.										
Apply the concept of a random variable to generate and interpret probability distribution including binomial, uniform, and chi square.										